SPOTLIGHT ON PHYSICAL SCIENCE

LIGHT

IT'S ENERGETIC

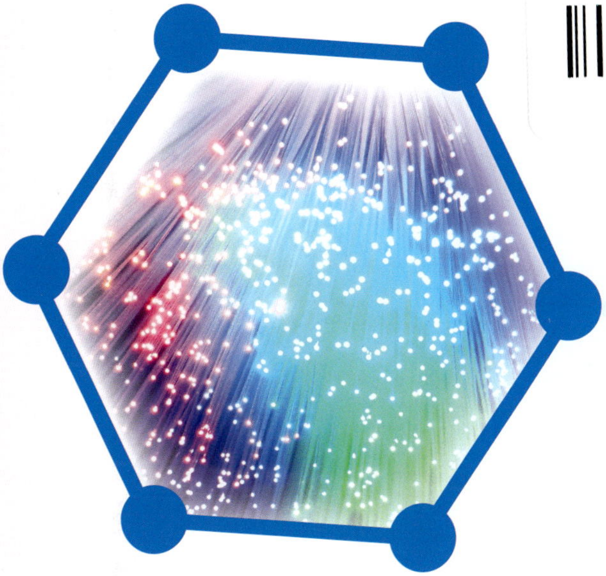

AMANDA VINK

PowerKiDS press

NEW YORK

Published in 2020 by The Rosen Publishing Group, Inc.
29 East 21st Street, New York, NY 10010

Editor: Elizabeth Krajnik
Book Design: Michael Flynn

Photo Credits: Cover, p. 1 asharkyu/Shutterstock.com; (series molecular background) pro500/Shutterstock.com; p. 5 (main) anko70/Shutterstock.com; p. 5 (inset) Maurice Tricatelle/Shutterstock.com; p. 6 Triff/Shutterstock.com; p. 7 https://commons.wikimedia.org/wiki/File:New_shot_of_Proxima_Centauri,_our_nearest_neighbour.jpg; p. 9 DE AGOSTINI PICTURE LIBRARY/De Agostini/Getty Images; p. 10 Everett Historical/Shutterstock.com; pp. 11, 17 Print Collector/Hulton Archive/Getty Images; p. 12 https://en.wikipedia.org/wiki/Christiaan_Huygens#/media/File:Christiaan_Huygens-painting.jpeg; p. 13 UniversalImagesGroup/Getty Images; p. 14 poliki/Shutterstock.com; pp. 15, 21 udaix/Shutterstock.com; p. 19 Fancy Tapis/Shutterstock.com; p. 20 Hogan Imaging/Shutterstock.com; p. 22 MANDY GODBEHEAR/Shutterstock.com; p. 23 hispan/Shutterstock.com; p. 25 Beloborod/Shutterstock.com; p. 27 Mykola Mazuryk/Shutterstock.com; p. 28 mansong suttakam/Shutterstock.com; p. 29 Komsan Loonprom/Shutterstock.com; p. 30 Dmitry Naumov/Shutterstock.com.

Cataloging-in-Publication Data

Names: Vink, Amanda.
Title: Light: it's energetic / Amanda Vink.
Description: New York : PowerKids Press, 2020. | Series: Spotlight on physical science | Includes glossary and index.
Identifiers: ISBN 9781725313194 (pbk.) | ISBN 9781725313224 (library bound) | ISBN 9781725313200 (6 pack)
Subjects: LCSH: Light--Juvenile literature. | Light and darkness--Juvenile literature.
Classification: LCC QC360.V56 2020 | DDC 535--dc23

Manufactured in the United States of America

CPSIA Compliance Information: Batch #CWPK20. For further information contact Rosen Publishing, New York, New York at 1-800-237-9932.

CONTENTS

CHAPTER ONE

BURN, BABY, BURN

Light affects everything on Earth. It allows humans and animals to see the world around them. Plants use light to make food and grow. Without light, Earth wouldn't be able to support life. Light is important, but what exactly is it? Where does light come from?

Light is a form of energy called radiant energy. Fire, light bulbs, and some electronics produce light. The universe has a fixed amount of energy. The law of **conservation** of energy states that energy in a closed system can't be created nor destroyed, but it can be changed from one form to another. A light bulb changes electrical energy into radiant energy. Plants turn radiant energy into chemical energy to make food through **photosynthesis**. When you eat food, whether it's plants or meat, you're putting chemical energy in the form of calories into your body. This cycle continues.

Some animals, such as fireflies and glowworms, make light through a chemical reaction in their bodies. This is called bioluminescence.

GLOWWORM

CHAPTER TWO

THE SUN

The sun is the star around which Earth and the other planets in our solar system revolve. Earth gets its light and heat from the sun. The sun is 93 million miles (149 million km) away from Earth. It's been active for more than 4.5 billion years.

The sun is a lot closer to us than the other stars in space are. Pictured here is our closest neighbor star, Proxima Centauri, which is 4.22 light-years away.

The sun creates light and heat through nuclear fusion, which is when the nuclei of atoms fuse to make new things. On the sun, fusion happens with atoms of **hydrogen**, which creates **helium** and releases enormous amounts of energy. The energy builds up and travels to the **photosphere** layer, and then out into the universe.

In a vacuum, light travels at a rate of 186,282 miles (299,792 km) per second. A light-year is the distance light travels in one Earth year, which is about 5.88 trillion miles (9.46 trillion km). It takes about 8.3 minutes for the sun's light to reach Earth.

CHAPTER THREE

AS FAST AS LIGHT!

Today we know that light is the fastest thing in the universe when it's in a vacuum. However, it took a long time for scientists to figure this out.

Many advances in the 17th century changed what scientists knew about light. Until then, many scientists thought light traveled from one place to another instantaneously, or instantly. In 1667, Italian astronomer Galileo Galilei tried to measure the speed of light. He had two people with covered lanterns stand on hills less than a mile apart. One person uncovered their lantern and when the second person saw the light, they uncovered their lantern too. This experiment wasn't very successful because the two people weren't standing far enough away from each other.

In 1676, Danish astronomer Ole Rømer successfully used an eclipse of one of Jupiter's moons to estimate the speed of light. His work determined that the speed of light isn't infinite, or doesn't go on forever.

Galileo Galilei didn't invent the telescope, but he did make many important improvements to it. He was the first person to look at the stars with a telescope.

CHAPTER FOUR

PARTICLE OR WAVE?

In the 17th century, scientists also developed the two main models of light: light as a particle and light as a wave.

In 1665, Italian **physicist** and mathematician Francesco Grimaldi's studies on light were published after his death. In his experiments, Grimaldi observed diffraction, or the slight bending of light around an object. This suggested that light didn't move in a straight line but rather as a wave.

A year later, English physicist Sir Isaac Newton used prisms, or pieces of glass with even angles and planes, to split white light into colors. The light shows up as individual colors because the colors all bend at a slightly different angle. Newton's work helped him come up with a theory: light was made up of particles that move in a straight line.

SIR ISAAC NEWTON

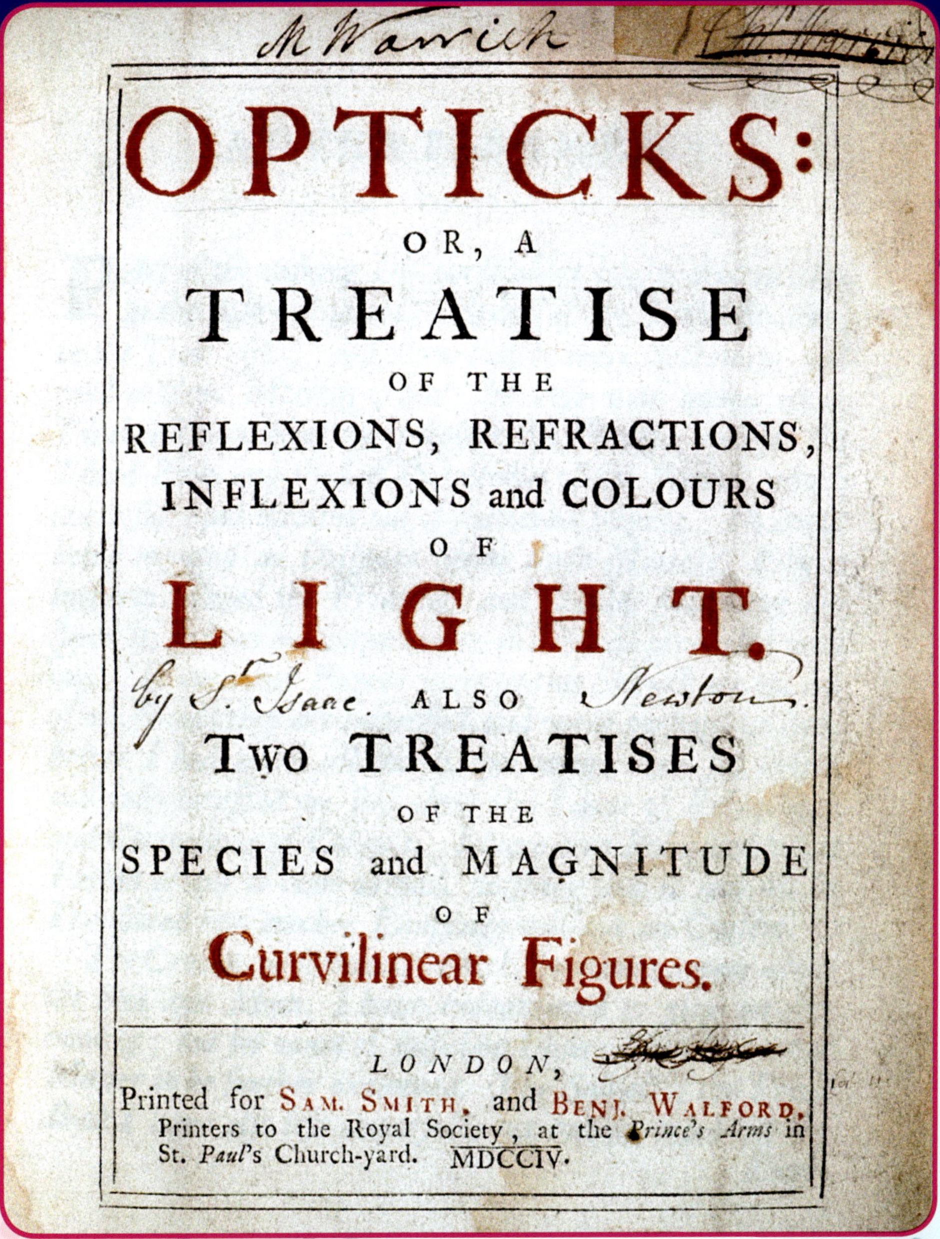

OPTICKS:

OR, A

TREATISE

OF THE

REFLEXIONS, REFRACTIONS, INFLEXIONS and COLOURS

OF

LIGHT.

ALSO

Two TREATISES

OF THE

SPECIES and MAGNITUDE

OF

Curvilinear Figures.

LONDON,

Printed for SAM. SMITH, and BENJ. WALFORD, Printers to the Royal Society, at the *Prince's Arms* in St. *Paul's* Church-yard. MDCCIV.

In 1704, Newton published *Opticks*, which detailed his studies in light, including the splitting of white light into colors.

When you drop a rock into calm water, ripples spread out in all directions. Many people once believed light moved in the same way.

In 1690, Dutch physicist Christiaan Huygens proposed that light behaves like a wave. He thought light moved through an invisible substance in the air he

CHRISTIAAN HUYGENS

Albert Einstein was working at an office and working on his doctorate in 1905 when he wrote a series of **innovative** papers, including one on his particle theory of light.

named “ether.” Eventually, scientists abandoned the idea of ether. However, Huygens’s wave theory did explain diffraction as well as some other properties of light. Until the mid-1800s, scientists were more convinced that light was a wave.

In 1905, German physicist Albert Einstein finished a scientific paper that described his particle theory of light. He argued that visible light exists as bundles of electromagnetic energy that would later be called **photons**. So really, light behaves like both a wave and a particle! Einstein’s theory was **radical** at the time and it wasn’t accepted for two decades.

CHAPTER FIVE

WHAT IS A PHOTON?

When Einstein wrote his paper in 1905, he didn't call the energy packets that transmit light "photons." American chemist Gilbert N. Lewis coined this term in 1926. Think of a light bulb. When it's on, it throws light in all directions. If you move farther from the light bulb, the light isn't as bright. At a molecular level, there's more space between the photons the farther away they are from the source of light.

PHOTOELECTRIC EFFECT

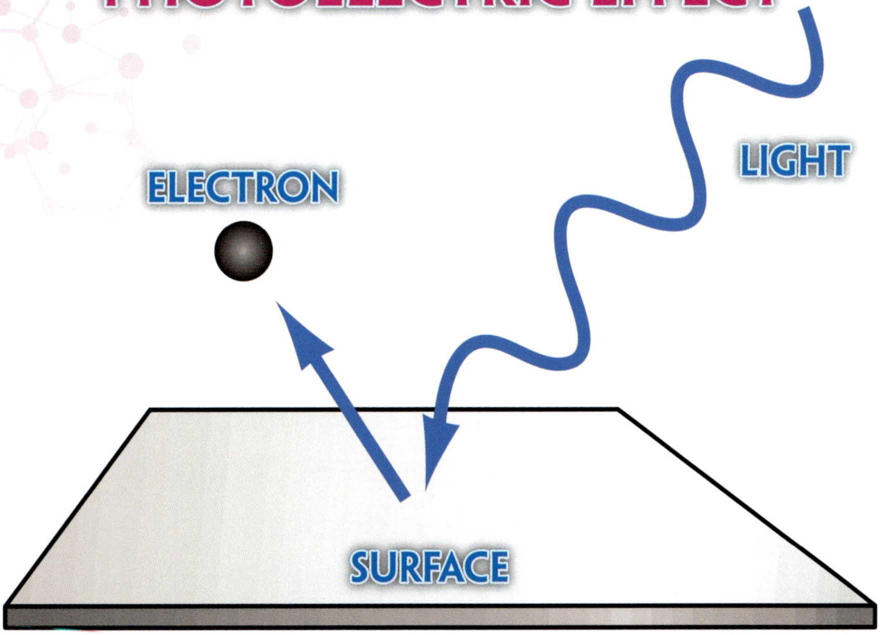

The photoelectric effect is when a metal plate ejects electrons when a light shines on it. The surface will only eject an electron when hit with the right **frequency** of light.

Photons are a type of subatomic particle, which means they're smaller than atoms. They're one of the fundamental components of all matter, and they're so small they can't be seen with the naked eye. Photons travel at the speed of light. Unlike many particles, photons don't have an electric charge. Photons don't have mass when they're at rest.

CHAPTER SIX

ELECTROMAGNETIC WAVES

Before Einstein proposed his theory, scientists had already figured out that light is made up of electromagnetic waves. In the 1860s, Scottish physicist James Clerk Maxwell combined the three fields of electricity, magnetism, and optics. He thought of the existence of electromagnetic **radiation**, and he also concluded that light itself was an electromagnetic wave. Electromagnetic waves don't need a medium, such as water, to travel through. Waves that need a medium to travel through, such as sound, are called mechanical waves.

Heinrich Hertz, a German physicist, applied Maxwell's conclusion to radio waves, and he discovered that radio waves are a form of light. In fact, scientists discovered there's more to light than just what we can see. Light is a type of electromagnetic energy and alternates between electric and magnetic fields.

Maxwell's electromagnetic theory influenced many physicists of the 20th century. His work is regarded as just as important as Newton's and Einstein's.

CHAPTER SEVEN

THE EM SPECTRUM

The electromagnetic **spectrum** includes different kinds of electromagnetic waves, including radio waves, microwaves, X-rays, and visible light rays. The highest part of a wave is called the crest. The lowest part is called the trough.

Waves in the EM spectrum can be measured in different ways. They can be measured in terms of their wavelengths, or the distance from one crest to the next. A single radio wave's wavelength can be very long, big enough to spread across a whole city! Meanwhile, gamma rays' wavelengths are extremely short. EM waves can also be measured in terms of their frequency, which is the number of waves that pass a point in one second. Frequency is measured in hertz (Hz). Amplitude is the height of a crest.

The EM spectrum includes all types of electromagnetic waves in the universe.

ELECTROMAGNETIC SPECTRUM

RADIATION TYPE

Radio | Microwave | Infrared | Visible | Ultraviolet | X-ray | Gamma ray

WAVELENGTH (M)

10^{3} | 10^{-2} | 10^{-5} | 0.5×10^{-6} | 10^{-8} | 10^{-10} | 10^{-12}

APPROXIMATE SCALE OF WAVELENGTH

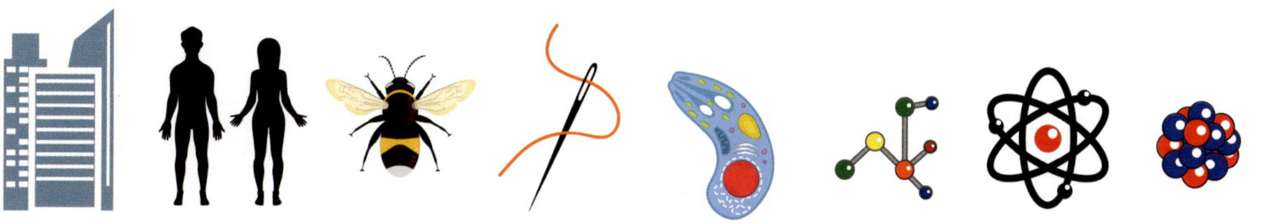

Buildings | Humans | Honeybee | Needle point | Protozoans | Molecules | Atoms | Atomic nuclei

FREQUENCY (Hz)

10^{4} | 10^{8} | 10^{12} | 10^{15} | 10^{16} | 10^{18} | 10^{20}

Each color of the visible light spectrum has a different wavelength. That's why we see them as distinct colors. The color red has the longest wavelength, measuring 70 nanometers, and violet has the shortest wavelength, measuring 380 nanometers. A nanometer is one-billionth of a meter.

The light a black T-shirt absorbs is most often converted into thermal, or heat, energy. That's why you feel hotter wearing a black T-shirt in the sun than if you were wearing a white T-shirt.

HOW WE SEE BLACK

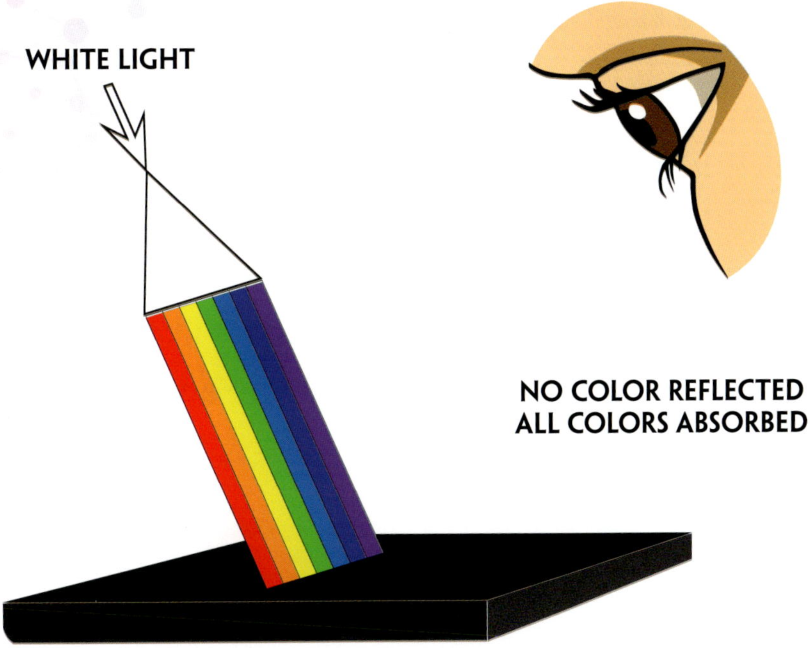

When light hits an object, a number of things can happen. First, the light may be absorbed and converted to heat. It may be transmitted, or go through the object. Light can also be reflected. Most matter reflects some light. This is how we see color. Light bounces off an object and is reflected into your eyes. When you see something red, that object is absorbing all colors on the visible light spectrum except red. White light is a combination of all the colors. Black is what you see when all the colors are absorbed.

CHAPTER EIGHT

THE HUMAN EYE

The human eye works much like a camera does. Light enters through the cornea, a clear layer that helps focus light and is like a camera lens. The iris controls how much light reaches the back of the eye. It does this by changing the size of the pupil, which is like a camera's aperture, or opening that controls the amount of light that passes through a lens.

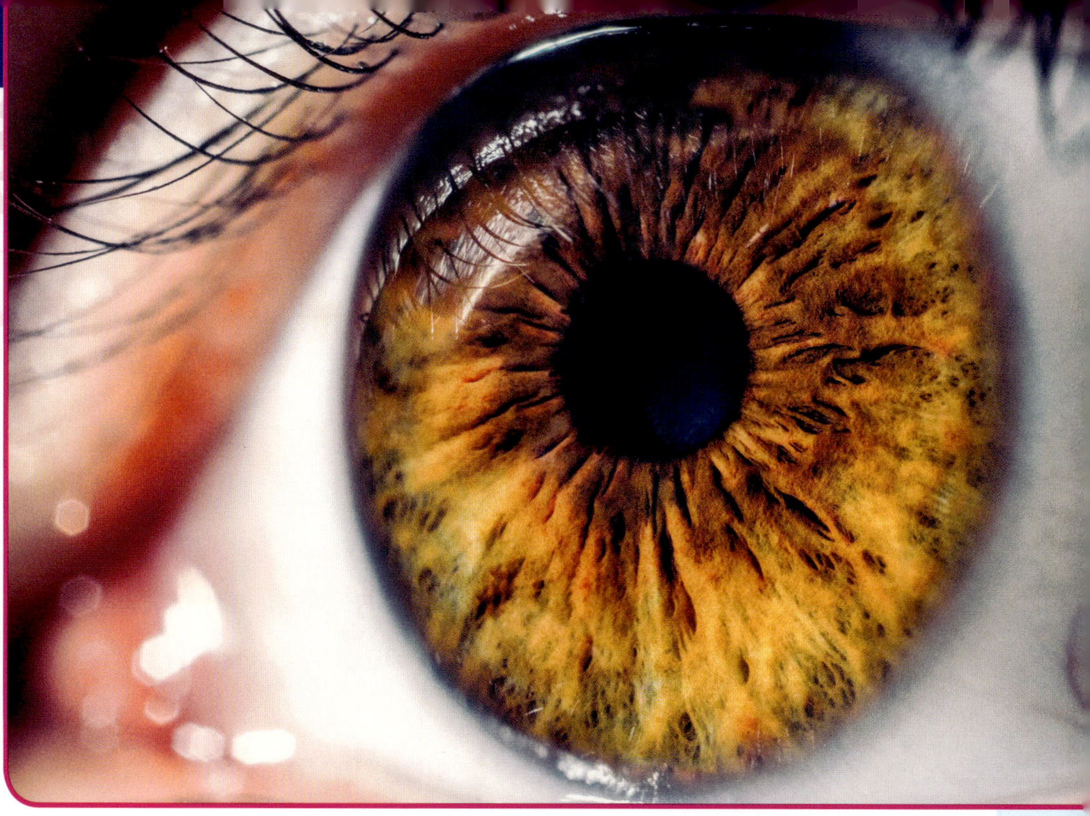

We don't actually see with our eyes. Our brain processes the light waves our eyes take in to create the images we see.

After the light passes the pupil, it hits the lens, which focuses light. From the lens, light reaches the retina, which changes what we see into electrical signals. The optic nerve, or thin part that controls movement and feeling by carrying messages between the brain and other parts of the body, carries these signals to the visual cortex, or the part of the brain that controls our sense of sight.

CHAPTER NINE

REFRACTION

When light passes through certain transparent objects, such as water or glass, it can be refracted. Refraction is when the direction of light is changed while the light passes through something. If you look at a straw through a glass of water, the straw looks broken or bent. This happens because the light's speed changed.

Light slows down when it travels through different materials. When white light is refracted, humans often see each of the colors of the visible light spectrum—red, orange, yellow, green, blue, indigo, and violet. In the case of a rainbow, light passes through droplets of water left over from a rainstorm. Remember how Newton used a prism to separate white light into different colors? The raindrops act like the prism in Newton's experiment.

The law of refraction, also called Snell's law, is named after 16th-century Dutch astronomer and mathematician Willebrord Snell.

CHAPTER TEN

REFLECTION

Have you ever looked up into the night sky at the moon? The moon doesn't make its own light. All the light coming from the moon is a reflection of light from the sun.

Some surfaces may interact with light in different ways. For example, light interacting with water or glass can be reflected, refracted, or pass right through without being affected all that much. When light bounces off a smooth surface, it reflects at the same angle with which it hit the surface. This is called specular reflection. If light bounces off an uneven surface, such as choppy water, it goes out in many different directions. This is called diffuse reflection. An **opaque** object blocks light. That's how we get shadows!

The moon only reflects between 3 percent and 12 percent of the sunlight that hits it. That's because it has an uneven surface.

CHAPTER ELEVEN

LIGHT AND TECHNOLOGY

Some animals can see different kinds of electromagnetic waves. Birds, bees, and some fish can see ultraviolet, or UV, waves, which have wavelengths that measure between 10 nanometers and 400 nanometers. Humans have developed machines to see light that isn't visible to the naked eye.

UV LIGHT

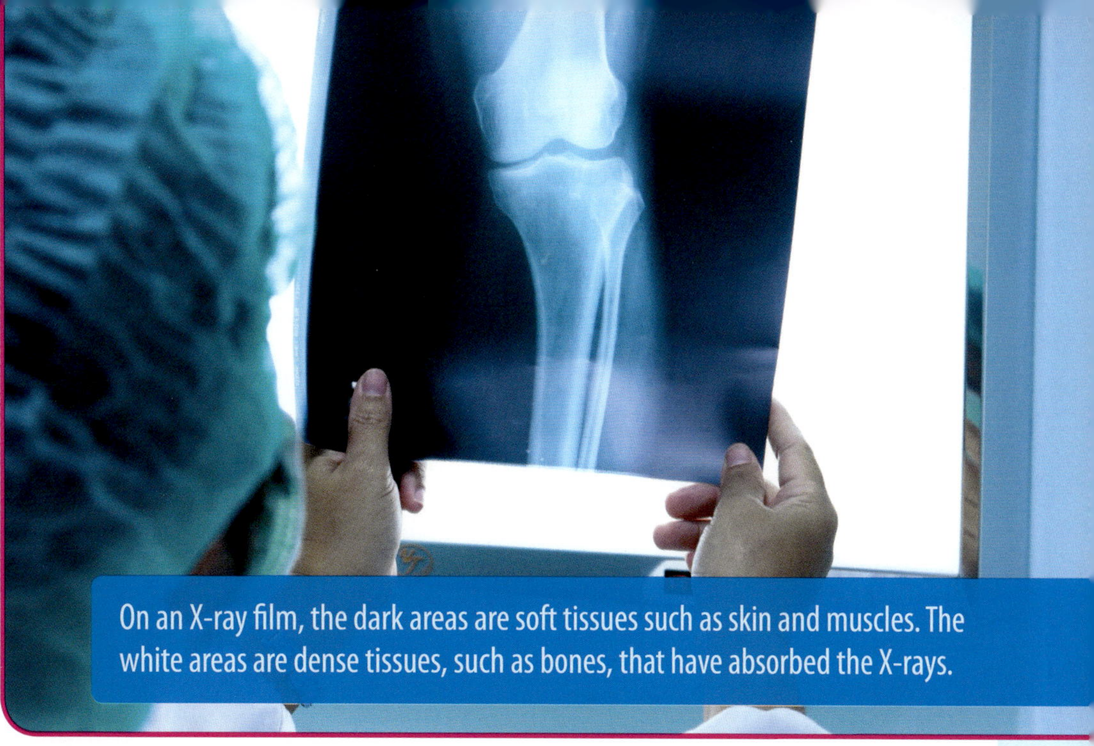

On an X-ray film, the dark areas are soft tissues such as skin and muscles. The white areas are dense tissues, such as bones, that have absorbed the X-rays.

Black lights emit, or give off, UV light. Substances called phosphors—which make up many things, including white clothes, our teeth, and our fingernails—convert the UV radiation they receive into visible light. This is why our teeth and white clothes glow under a black light!

An X-ray machine works by sending out a beam of electrons. Materials that are closely packed, such as our bones, absorb the radiation, while the X-rays go through softer body tissues. The beam of electrons makes an image on a metal film. This film is held up to a light to show doctors things such as broken bones.

CHAPTER TWELVE

LIGHT AND YOU

The study of light has led to many important scientific discoveries and inventions. For example, humans send information using light through optic cables. We now collect the sun's radiation and turn it into energy by using photovoltaic cells on solar panels. We cook food in microwaves, which use the shortest radio waves. Learning about light has allowed us to create telescopes that can see much farther into space than ever imagined. Even eyeglasses are a result of the study of light!

Learning how light works has helped scientists learn a lot about the way the world works, too. We know that light and heat from the sun affect the length of our days and our weather. Take a look around you. How is light affecting what you see right now?

GLOSSARY

conservation (kahn-suhr-VAY-shuhn) The preservation of a physical quantity during transformations or reactions.

frequency (FREE-kwuhn-see) The rate of repetition.

helium (HEE-lee-uhm) A very light gas and chemical element that is found in various natural gases, will not burn, and is used in balloons.

hydrogen (HY-druh-juhn) A colorless, odorless, and tasteless flammable gas that is the lightest of the chemical elements.

innovative (IH-nuh-vay-tihv) Using or showing new methods or ideas.

opaque (oh-PAYK) Not letting light through; not transparent.

photon (FOH-tahn) A tiny particle of light or electromagnetic radiation.

photosphere (FOH-tuh-sfeer) The bright surface layer of a star.

photosynthesis (foh-toh-SIHN-thuh-suhs) The process by which green plants use light to turn water and carbon dioxide into food.

physicist (FIH-zuh-sihst) Someone who studies physics, or the science that deals with matter and energy and their interactions.

radiation (ray-dee-AY-shuhn) The process of giving off energy in the form of waves or particles.

radical (RAA-dih-kuhl) Very new and different, not ordinary.

spectrum (SPEHK-truhm) A continuous sequence or range.

INDEX

PRIMARY SOURCE LIST

Page 7
Proxima Centauri. Photograph. Taken by the Hubble telescope. October 28, 2013. Now kept by ESA/Hubble Images.

Page 9
Telescope. Created in the 17th century. Galileo Galilei. Museo Della Specola Universita Di Bologna Dipartimento Di Astronomia, Italy.

Page 13
Albert Einstein. Photograph. ca. 1905. Now kept by Universal Images Group/Getty Images.

WEBSITES

Due to the changing nature of Internet links, PowerKids Press has developed an online list of websites related to the subject of this book. This site is updated regularly. Please use this link to access the list: www.powerkidslinks.com/SOPS/light